Birds of Prey

KITES

JANET DAVIS-CASTRO

BLACK
RABBIT
BOOKS

BOLT

Bolt is published by Black Rabbit Books
P.O. Box 227, Mankato, Minnesota, 56002
www.blackrabbitbooks.com
Copyright © 2023 Black Rabbit Books

Marysa Storm, editor; Michael Sellner, designer and
photo researcher

Library of Congress Cataloging-in-Publication Data
Names: Davis-Castro, Janet, author.
Title: Kites / by Janet Davis-Castro.
Description: Mankato, Minnesota : Black Rabbit Books, [2023] | Series:
    Bolt. birds of prey | Includes bibliographical references and index. |
    Audience: Ages 8-12 | Audience: Grades 4-6 | Summary: "Equipped with
    strong senses and graceful features, kites rule the skies. They also
    capture the imagination of readers everywhere. With an engaging design
    focused around expertly leveled text and colorful infographics that
    support visual literacy, introduce reluctant and beginning readers to
    the life cycle, features, and diets of these powerful birds of prey"
    Provided by publisher.
Identifiers: LCCN 2020017983 (print) | LCCN 2020017984 (ebook) |
    ISBN 9781623105556 (hardcover) | ISBN 9781644664865 (paperback) |
    ISBN 9781623105617 (ebook)
Subjects: LCSH: Red kite–Juvenile literature.
Classification: LCC QL696.F32 D38 2023 (print) | LCC QL696.F32 (ebook) |
    DDC 598.9-dc23
LC record available at https://lccn.loc.gov/2020017983
LC ebook record available at https://lccn.loc.gov/2020017984

Contents

CHAPTER 1

FLASHY
Fliers

A kite glides in circles. The bird moves its tail feathers to twist and turn. It's searching for food flying below. The flashy flier spots a wasp. In a smooth movement, the bird swoops down. It then turns upside down and dives backward. It grabs its **prey** with its beak and eats the meal in midair.

COMPARING WINGSPANS

MISSISSIPPI
KITE

29 to 33 INCHES
(74 to 84 cm)

Different Types

Kites are a type of small hawk. These graceful birds of prey have long wings and short beaks. There are many types of kites. They come in different sizes. Their wingspans range from about 30 to almost 80 inches (76 to 203 centimeters). Some use their wings to soar through the sky. Others flap them quickly to hover above the ground.

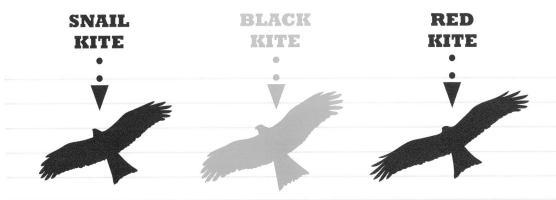

SNAIL KITE

BLACK KITE

RED KITE

42 to 48 INCHES
(107 to 122 cm)

55 to 59 INCHES
(140 to 150 cm)

69 to 77 INCHES
(175 to 196 cm)

PARTS OF A KITE

EYES

WINGS

BEAK

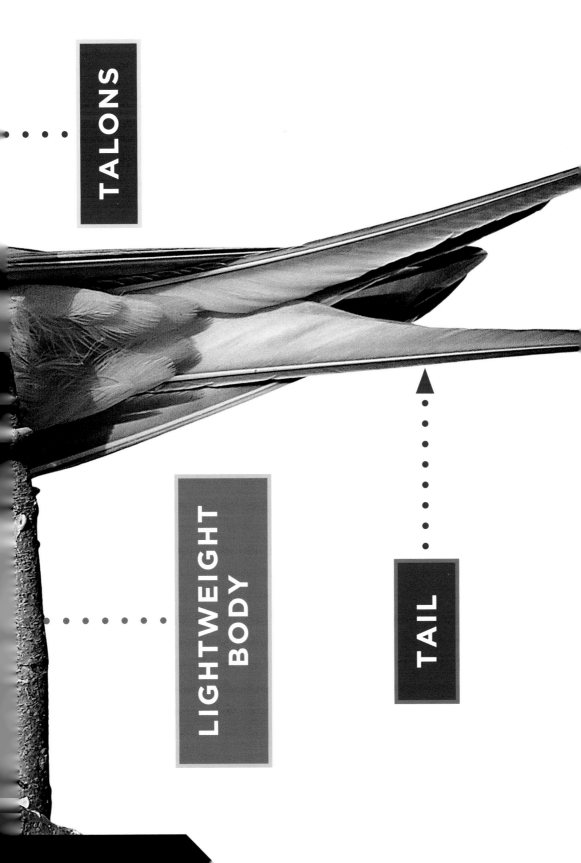

TALONS

LIGHTWEIGHT
BODY

TAIL

CHAPTER 2

At HOME and on the Hunt

These birds spend their days hunting. They often feast on **reptiles**, worms, bugs, and small **mammals**. Some eat eggs or dead animals they find.

Kites' features help them hunt. Good eyesight helps them see prey from the sky. Light bodies and long wings help them glide for a long time. Gliding saves energy. Kites' tails work like steering wheels. They help the birds change direction quickly to chase after prey.

To hunt, white-tailed kites often hover like helicopters. They drop down on their prey with powerful talons.

Hunting

Not all kite beaks and talons are the same. They use them to hunt in different ways. For example, snail kites have long talons. They use them to pluck snails from the water. The kites' curved beaks then pull the snails from their shells. Other kites have strong talons. They use them to attack and tear apart animals. • • • •

Habitats

Kites live in warm spots all over the world. They fly above grasslands, wetlands, forests, and near rivers. Some kites migrate. Before winter, they fly thousands of miles to warmer spots. They travel the same path back home during the spring.

Most swallow-tailed kites live in Florida. Each winter, they travel to Brazil. In spring, they fly back to Florida.

Florida

Brazil

KITE RANGE MAP

Kites live on every continent except Antarctica.

NORTH
AMERICA

SOUTH
AMERICA

EUROPE

ASIA

AFRICA

AUSTRALIA

ANTARCTICA

CHAPTER 3

KITE Families

Kites have families made up of parents and their chicks. To start a family, kites look for **mates**. They fly and chase each other to show off their **skills**. Females choose the best fliers.

After mating, most pairs build nests high in trees. They work together to build the nests from twigs and grasses.

Some kites build their nests next to each other. They create kite neighborhoods.

Nest Life

EGGS

BROODING

FEEDING • • •

Babies

Once the nest is ready, females lay one to five eggs. Usually, females keep the eggs warm while males bring them food. The eggs hatch after four to five weeks. Both parents protect and feed the hatchlings. After several weeks, the young kites learn to fly.

COMPARING YOUNG

Average Clutch Size

black kite	Mississippi kite	swallow-tailed kite	white-tailed kite

2   **2**

2–3

4